Photosynthesis

by Grace Hansen

Abdo Kids
BEGINNING SCIENCE: ECOLOGY

Abdo Kids Jumbo is an Imprint of Abdo Kids
abdobooks.com

abdobooks.com

Published by Abdo Kids, a division of ABDO, P.O. Box 398166, Minneapolis, Minnesota 55439.

Printed in the United States of America, North Mankato, Minnesota.

102019

012020

THIS BOOK CONTAINS RECYCLED MATERIALS

Photo Credits: iStock, Science Source, Shutterstock

Production Contributors: Teddy Borth, Jennie Forsberg, Grace Hansen
Design Contributors: Dorothy Toth, Pakou Moua

Library of Congress Control Number: 2019941256

Publisher's Cataloging-in-Publication Data

Names: Hansen, Grace, author.

Title: Photosynthesis / by Grace Hansen

Description: Minneapolis, Minnesota : Abdo Kids, 2020 | Series: Beginning science: ecology | Includes online resources and index.

Identifiers: ISBN 9781532188961 (lib. bdg.) | ISBN 9781644942697 (pbk.) | ISBN 9781532189456 (ebook) | ISBN 9781098200435 (Read-to-Me ebook)

Subjects: LCSH: Photosynthesis--Juvenile literature. | Plants--Juvenile literature. | Food webs (Ecology)--Juvenile literature. | Ecology--Juvenile literature. | Light--Juvenile literature.

Classification: DDC 581.13--dc23

Table of Contents

What Is Photosynthesis?

Every day, important things happen on Earth. One of these things is called photosynthesis. Without it, life would not be possible.

Photosynthesis is a process carried out by green plants. With it, plants make their own food.

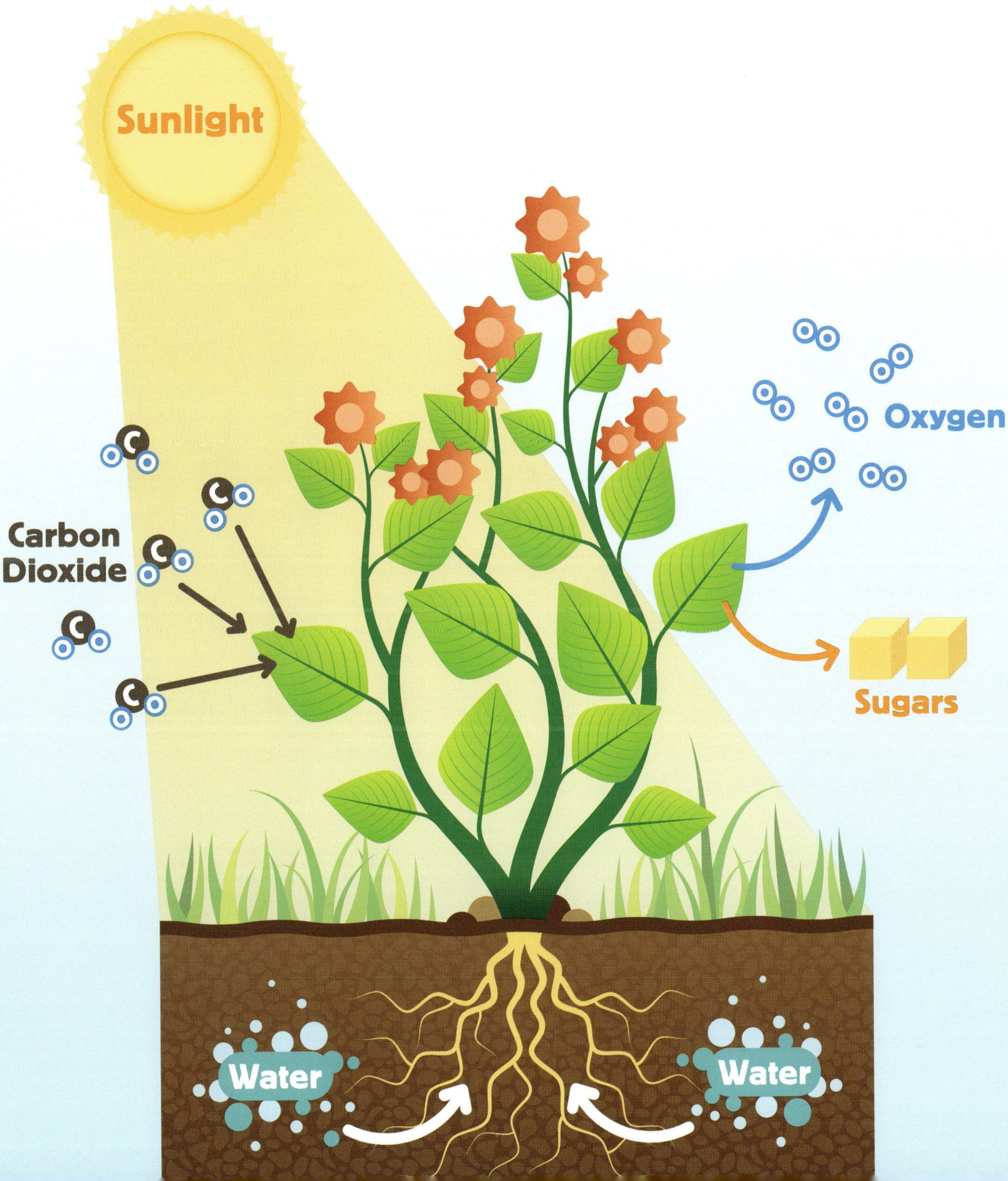
Sunlight
Carbon
Dioxide
Oxygen
Sugars
Water
Water

How It Works

Plants are green because of a chemical called **chlorophyll**. When sunlight hits green leaves, chlorophyll captures it.

chlorophyll

Plants also absorb a gas from the air called carbon dioxide. They take in water through their **roots**.

All of these things are needed for the plant to make sugar. Sugar is energy for the plant. The plant uses this energy to grow and fuel itself.

During this process, plants also release oxygen into the air. Humans and animals need oxygen to breathe!

O_2
O_2
O_2
CO_2
CO_2
CO_2

Photosynthesis in The Food Chain

Food chains show how energy flows through an **ecosystem**. Earth's main source of energy is the sun. Energy from the sun passes onto plants.

Plants pass their energy onto the animals that eat them. The energy keeps flowing when other animals eat the plant-eating animals.

When animals and plants die, **decomposers** break down their bodies. The nutrients from their bodies go back into the soil. None of this would be possible without photosynthesis.

Let's Review!

- Photosynthesis makes life on Earth possible.
- Photosynthesis is carried out by green plants. They use this process to make their own food.
- Green plants need sunlight, water, and carbon dioxide to carry out photosynthesis.
- Sugar and oxygen are both made through photosynthesis.
- Sunlight is a form of energy. That energy passes onto plants. Plants convert that energy into sugar. Humans and animals get energy from eating plants.

Glossary

chlorophyll – the green matter in the leaves and stems of plants that is necessary to produce food for plants from sunlight.

decomposer – an organism, like bacteria or fungus, that breaks down dead, organic matter in the environment, sometimes after a scavenger is done with it.

ecosystem – a community of livings things, together with their environment.

root – the part of a plant that usually grows underground. Roots take up water and nutrients and hold the plant in the soil.

Index